Building the Future:
Uses of AI in Construction

I0417250

Devon Pinard

Table of Contents

Page 4 – Introduction to Artificial Intelligence in Construction

Page 7 – Benefits and Challenges of Implementing AI in Construction

Page 10 – The Role of Mahine Learning in Project Planning and Scheduling

Page 13 – Use of Robotics and Automation in Construction Processes

Page 16 – AI Applications for Safety and Risk Management in Construction

Page 19 – Enhancing Building Design and Simulation with AI Technology

Page 22 – Intelligent Decision-Making with Advanced Data Analytics

Page 25 – Implementing AI for Predictive Maintenance and Asset Management

Page 28 – Smart Construction Site Monitoring and Progress Tracking

Page 31 – Future Trends and Innovations in AI for Construction Industry

Introduction to Artificial Intelligence in Construction

Artificial intelligence (AI) is redefining the construction industry landscape by leveraging advanced technologies to enhance efficiency, productivity, and safety across project lifecycles. Drawing on machine learning, robotics, and data analytics, AI is revolutionizing traditional construction practices and setting new benchmarks for project management and delivery.

Within the realm of project planning and scheduling, AI is driving significant advancements through automated tools that optimize resource allocation, streamline workflows, and mitigate project delays. By analyzing intricate project data and identifying potential bottlenecks, AI scheduling algorithms empower construction teams to make informed decisions that improve project timelines and overall performance.

Beyond scheduling, AI is playing a critical role in predictive maintenance strategies within construction. By integrating IoT sensors and AI algorithms, companies can monitor equipment and infrastructure health in real-time, enabling proactive maintenance and reducing costly downtime. This predictive maintenance approach not only enhances operational efficiency but also bolsters workplace safety by addressing maintenance needs promptly.

AI is also revolutionizing quality control practices in construction by harnessing computer vision technology and AI-powered drones for automated inspections. These technologies enable construction teams to detect defects and deviations from design specifications with greater accuracy and speed, ensuring that projects meet regulatory standards and client expectations. By identifying quality issues early on, AI-driven quality control measures enhance overall project quality and reduce rework costs.

In the realm of safety monitoring, AI is taking safety practices in construction to new heights by employing advanced safety solutions. AI-powered tools can analyze on-site conditions, identify potential hazards, and provide real-time alerts to prevent accidents and injuries. By integrating AI-driven safety measures, construction companies can cultivate a safer work environment, safeguarding the well-being of their employees and promoting a culture of safety on construction sites.

The adoption of AI technologies in construction signifies a change in thinking toward innovation, efficiency, and success in the industry. As construction firms continue to embrace AI-driven solutions, they will unlock new opportunities for growth, accuracy, and competitiveness in their projects, ultimately shaping a future where smart technologies optimize project outcomes and redefine the built environment.

Benefits and Challenges of Implementing AI in Construction

In addition to improving project management processes and enhancing safety on construction sites, Artificial Intelligence (AI) technologies offer a range of advanced capabilities that can revolutionize the way construction projects are planned, designed, and executed. One key area where AI is making significant strides is in the field of Building Information Modeling (BIM). BIM is a 3D modeling process that allows architects, engineers, and contractors to collaborate on a digital representation of a building before it is constructed. By integrating AI into BIM systems, construction teams can optimize design decisions, reduce errors, and improve collaboration across disciplines.

AI-enabled algorithms can analyze BIM data to identify potential clashes or conflicts in the design, such as overlapping structural elements or incompatible building systems. By detecting these issues early in the design phase, AI can help prevent costly rework during construction and ensure that the final building meets project requirements. Furthermore, AI can assist construction teams in generating optimized design alternatives based on performance criteria such as cost, energy efficiency, and sustainability. This enables architects and engineers to explore a wider range of design options and make informed decisions that align with project goals.

Another area where AI is driving innovation in construction is in the realm of robotic automation. Robotics technology combined with AI capabilities allows for the automation of repetitive construction tasks, such as bricklaying, welding, and concrete pouring. By deploying robotic systems on construction sites, companies can increase efficiency, accelerate project timelines, and reduce labor costs. For example, robotic bricklaying machines can lay bricks at a faster pace than human workers, enhancing productivity while maintaining precision and quality. Similarly, robotic exoskeletons can assist construction workers in lifting heavy materials, reducing the risk of injuries and improving overall job site safety.

Moreover, AI-powered predictive analytics is revolutionizing how construction companies manage and maintain infrastructure assets. By analyzing historical project data and real-time sensor data from building systems, AI algorithms can predict potential equipment failures, optimize maintenance schedules, and extend the lifespan of critical infrastructure. This predictive maintenance approach minimizes downtime, reduces operational costs, and enhances the reliability of construction assets.

As AI technologies continue to mature and evolve, the construction industry stands to benefit from increased efficiency, improved safety, and enhanced project outcomes. By embracing AI as a strategic tool for innovation and progress, construction companies can stay competitive in a rapidly changing market landscape and drive sustainable growth in the built environment. The integration of AI in construction represents a new era of possibilities, where smart technologies empower industry professionals to reshape how buildings are designed, built, and maintained in the digital age.

The Role of Machine Learning in Project Planning and Scheduling

In the construction industry, project planning and scheduling are crucial aspects that can significantly impact the success of a project. With the ever-increasing complexity and scale of construction projects, efficient planning and scheduling are essential to ensure on-time delivery, cost-effectiveness, and quality outcomes.

Machine learning, a subset of artificial intelligence, has emerged as a powerful tool that has the potential to revolutionize how project planning and scheduling are conducted in the construction sector. By leveraging historical project data, machine learning algorithms can analyze patterns, identify correlations, and make predictions that optimize project timelines and resource allocation.

One of the key advantages of using machine learning in project planning is the creation of predictive models based on historical data. These models can forecast project timelines, anticipate potential risks and delays, and recommend strategies to mitigate them. By learning from past project outcomes, machine learning algorithms enhance decision-making processes and improve overall project performance.

Furthermore, machine learning can play a pivotal role in optimizing resource allocation within construction projects. By analyzing data on resource availability, utilization rates, and project requirements, machine learning models can suggest optimal resource allocation strategies to maximize productivity and minimize costs. This data-driven approach empowers construction professionals to make well-informed decisions and allocate resources more effectively.

An additional critical application of machine learning in construction project planning is the identification of potential project bottlenecks and inefficiencies. By analyzing various project parameters and historical data, machine learning algorithms can pinpoint areas of concern that may hinder project progress. This initiative-taking approach enables project managers to address issues promptly, implement necessary adjustments, and uphold project timelines.

The integration of machine learning in project planning and scheduling offers immense potential for enhancing efficiency, productivity, and project outcomes in the construction industry. By harnessing the power of data analytics and artificial intelligence, construction professionals can drive innovation, streamline processes, and deliver successful projects that exceed client expectations.

Use of Robotics and Automation in Construction Processes

In the construction industry, the integration of robotics and automation has become a pivotal force reshaping traditional building practices and revolutionizing the way structures are designed, built, and maintained. This technological transformation represents a change in thinking that transcends mere trendiness, fundamentally altering the landscape of construction by introducing new efficiencies and possibilities.

One of the key advantages of leveraging robotics in construction processes lies in the automation of labor-intensive tasks that have historically been time-consuming and prone to human error. By employing robots equipped with advanced sensors and algorithms, construction companies can achieve unparalleled precision and consistency in tasks such as bricklaying, concrete pouring, and structural assembly. The use of robotic arms with specialized end-effectors not only ensures accuracy but also enhances overall construction quality by minimizing deviations from the intended design.

Beyond increasing efficiency and quality, the adoption of robotics and automation in construction also presents a significant opportunity for improving job site safety. Construction sites are inherently hazardous environments, characterized by a variety of risks that can pose threats to workers' well-being. By delegating high-risk tasks to robots, such as working at heights or handling heavy materials, human workers can be safeguarded from potential accidents and injuries. This shift toward a safer work environment is a crucial benefit of incorporating automation technologies in construction practices.

Moreover, the incorporation of automation technologies such as drones and autonomous vehicles has brought about a new era of innovation in the industry. Drones equipped with high-resolution cameras and LiDAR sensors enable real-time aerial surveys, site mapping, and progress monitoring, providing project managers with valuable insights and data for decision-making. Similarly, autonomous vehicles equipped with advanced navigation systems optimize material transportation within construction sites, reducing manual labor and enhancing overall coordination efficiency.

Embracing robotics and automation is not merely a matter of staying competitive in the construction industry; it is a strategic imperative for companies seeking to thrive in a rapidly evolving technological landscape. While challenges may arise during the initial adoption phase, the long-term benefits in terms of cost savings, improved safety, and accelerated project delivery far outweigh the initial hurdles. By embracing these transformative technologies, construction firms can position themselves at the forefront of innovation and set new standards for excellence in the built environment.

AI Applications for Safety and Risk Management in Construction

In the construction industry, safety and risk management play crucial roles in ensuring the successful completion of projects. As projects become more complex and larger in scale, the need for advanced technologies to enhance safety practices and mitigate risks becomes increasingly important. Artificial intelligence (AI) has emerged as a powerful tool in addressing these challenges by revolutionizing safety management in the construction sector.

AI's impact on safety management is particularly evident in the application of computer vision technology at construction sites. By utilizing AI-enabled cameras, project managers can access real-time data to monitor workers' activities, identify potential hazards, and detect safety violations. The implementation of computer vision technology allows for immediate intervention when safety issues arise, thus ensuring a safer working environment for everyone involved on the construction site.

Furthermore, AI algorithms can analyze vast amounts of data to predict and mitigate risks on construction projects. By leveraging predictive analytics, stakeholders can identify patterns and trends that may lead to safety hazards or project delays. This data-driven approach empowers project teams to take proactive measures, allocate resources efficiently, and address potential risks before they escalate, enhancing safety outcomes and project success.

AI technology also holds the potential to optimize safety training programs for construction workers. By analyzing individual performance data and feedback, AI systems can tailor training modules to address specific safety concerns and improve workers' adherence to safety protocols. This personalized training approach not only enhances workers' understanding of safety practices but also cultivates a culture of safety within the organization, leading to improved safety outcomes.

Moreover, the integration of AI in safety and risk management in construction drives continuous improvement and innovation. By leveraging AI-powered tools, stakeholders gain valuable insights, streamline decision-making processes, and adapt quickly to changing conditions on-site. This dynamic approach not only enhances safety practices but also improves overall project efficiency and performance, ultimately leading to successful project outcomes.

In conclusion, the adoption of AI in safety and risk management in construction marks a significant shift in the industry's approach to safety practices. Embracing AI technologies as strategic tools enables construction stakeholders to proactively manage risks, enhance safety practices, and deliver successful projects that meet quality, schedule, and budget requirements. The incorporation of AI in safety management fosters innovation, promotes a culture of safety, and ensures the long-term success of construction projects.

Enhancing Building Design and Simulation with AI Technology

In recent years, the use of artificial intelligence (AI) technology has revolutionized the way building design and simulation are conducted in the construction industry. Through the implementation of AI algorithms and machine learning models, architects and engineers can optimize the design process, improve energy efficiency, and enhance overall building performance.

One key aspect of enhancing building design with AI technology is the utilization of generative design tools. These tools enable designers to input specific parameters and constraints, allowing the AI system to generate a wide range of design options that meet the project requirements. This iterative process can result in innovative and efficient design solutions that may not have been considered without the help of AI.

Generative design tools work by employing algorithms that explore a vast range of design possibilities, making decisions based on performance objectives and constraints set by the designer. By analyzing massive amounts of data and iterating quickly, AI systems can produce designs that are optimized for several factors such as material usage, structural integrity, environmental impact, and aesthetic appeal. This approach not only saves time but also leads to more sustainable and resource-efficient building solutions.

Additionally, AI technology can be used to simulate and analyze various aspects of a building's performance, such as thermal dynamics, lighting conditions, and structural integrity. By inputting data and running simulations, designers can quickly assess unique design alternatives and make informed decisions based on the results provided by the AI system.

One prominent application of AI in building design is in energy modeling. By using AI algorithms to analyze energy consumption patterns and building performance data, designers can optimize building designs to minimize energy usage and reduce environmental impact. AI tools can predict energy demands, recommend energy-efficient solutions, and help track and control energy usage during the building's operation phase.

Furthermore, AI technology is increasingly being used for predictive maintenance in buildings. By analyzing data from sensors and building management systems, AI algorithms can detect patterns indicative of potential equipment failures or maintenance needs. This proactive approach to maintenance not only reduces downtime and repair costs but also improves overall building performance and occupant comfort.

Moreover, AI-driven building design is paving the way for more sustainable and resilient buildings. Through advanced data analytics and machine learning, designers can assess the environmental impact of unique design choices, allowing for the creation of structures that are not only energy-efficient but also environmentally friendly. By optimizing building performance through AI, we can create spaces that are healthier for occupants and have a reduced carbon footprint.

In conclusion, the integration of AI technology in building design and simulation is revolutionizing the construction industry by enabling designers and engineers to create more sustainable, efficient, and high-performing buildings. From generative design tools to energy modeling and predictive maintenance, AI is reshaping the way buildings are conceptualized, constructed, and maintained, leading to a more innovative and environmentally conscious built environment.

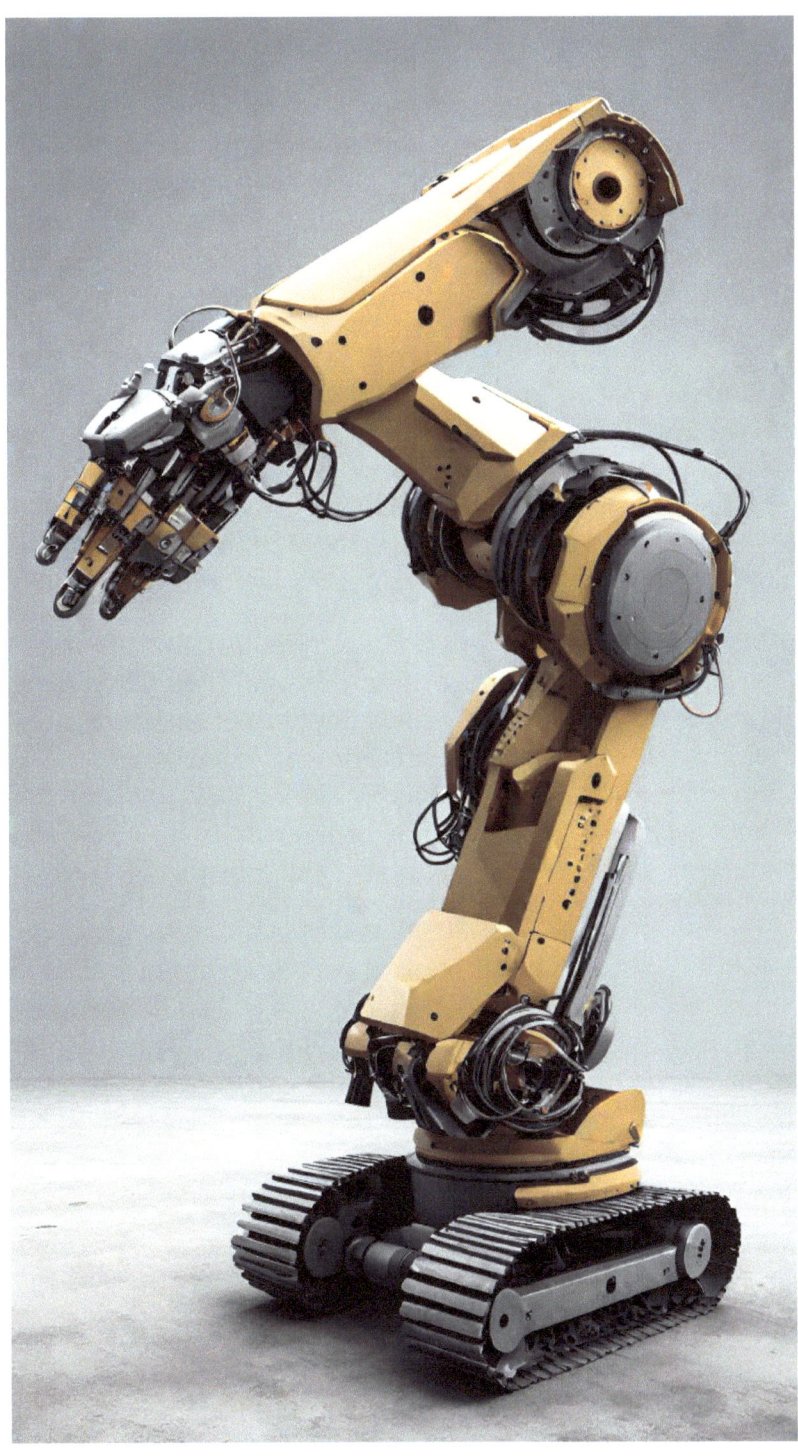

Intelligent Decision-Making with Advanced Data Analytics

In the realm of intelligent decision-making with advanced data analytics in the construction industry, there exists a wealth of opportunities for firms to transform their operations and drive better outcomes. By embracing cutting-edge technologies such as artificial intelligence, machine learning, and predictive analytics, construction companies can unlock new levels of efficiency, productivity, and innovation.

One key area where data analytics can make a significant impact is in the realm of project risk management. By leveraging historical data, project managers can identify patterns and trends that signal potential risks or delays in their construction projects. This approach allows firms to take preemptive actions to mitigate these risks and ensure that projects stay on track and within budget.

Moreover, data analytics can also revolutionize the way companies manage their supply chains. By analyzing data on material availability, pricing fluctuations, and supplier performance, construction firms can optimize their procurement processes, reduce costs, and enhance their overall supply chain efficiency. This data-driven approach enables companies to make more informed decisions about sourcing, inventory management, and logistics, leading to greater operational success.

Additionally, the integration of data analytics into construction project management can revolutionize scheduling and resource allocation. By utilizing predictive analytics models, firms can optimize their project schedules, allocate resources more effectively, and identify potential bottlenecks before they occur. This initiative-taking approach not only improves project efficiency but also enhances workforce productivity and reduces the likelihood of costly delays.

Furthermore, data analytics can play a pivotal role in enhancing collaboration and communication within construction teams. By centralizing project data in a digital platform and providing real-time insights to all stakeholders, firms can foster greater transparency, alignment, and decision-making across different departments and project phases. This data-driven collaboration empowers teams to work more cohesively, make better-informed decisions, and ultimately deliver superior project outcomes.

The adoption of data analytics in construction can also lead to enhanced safety measures on construction sites. Through the analysis of historical safety incident data, construction companies can identify patterns that may indicate potential hazards or risks to worker safety. By implementing predictive analytics tools, firms can proactively address safety concerns, implement preventive measures, and create a safer working environment for their employees.

Moreover, data analytics can drive sustainability initiatives within the construction industry. By analyzing environmental impact data and resource consumption patterns, firms can identify opportunities to reduce waste, optimize energy usage, and minimize their ecological footprint. This data-driven approach enables companies to make more sustainable decisions throughout the project lifecycle, from design and construction to operation and maintenance, aligning with global efforts to promote environmental sustainability in the built environment.

In conclusion, the integration of advanced data analytics into construction operations presents a myriad of opportunities for firms to elevate their performance and competitiveness. By harnessing the power of data-driven decision-making, construction companies can navigate challenges more effectively, optimize their processes, and achieve greater success in an increasingly competitive market landscape.

Implementing AI for Predictive Maintenance and Asset Management

Implementing artificial intelligence (AI) for predictive maintenance and asset management in the construction industry has ushered in a new era of operational efficiency and cost-effectiveness. AI algorithms, powered by machine learning and data analytics, have the capacity to revolutionize how maintenance practices are conducted and assets are managed in construction projects.

The shift from traditional reactive maintenance strategies to initiative-taking predictive maintenance is a significant advancement enabled by AI technology. By analyzing large datasets of historical equipment performance and failure patterns, AI systems can forecast impending issues before they arise. This predictive capability allows maintenance teams to intervene strategically, preventing unexpected breakdowns and minimizing downtime on construction sites. The ability to anticipate maintenance needs based on data-driven insights enables construction companies to optimize resource allocation and enhance overall project efficiency.

Real-time monitoring of equipment health through AI-enabled sensors and predictive analytics provides a continuous stream of valuable data for maintenance decision-making. By tracking key metrics such as temperature, vibration levels, and operational parameters, AI systems can detect anomalies and irregularities that may signal equipment malfunctions. This initiative-taking monitoring approach enables maintenance teams to address potential problems in a timely manner, ensuring that equipment operates at peak performance levels and reducing the likelihood of costly repairs or replacements.

In addition to predictive maintenance, AI technology offers transformative benefits for asset management within the construction industry. By analyzing data on asset utilization, condition, and maintenance history, AI systems can optimize asset lifecycle management strategies. This data-driven approach allows construction companies to make informed decisions about asset investments, maintenance schedules, and replacement cycles, leading to improved asset performance and longevity.

Furthermore, AI-powered predictive maintenance enhances safety in construction operations by identifying potential hazards and maintenance requirements in real time. By flagging critical issues and recommending preventive measures, AI systems contribute to creating a safer work environment for construction personnel. The initiative-taking maintenance actions driven by AI technology help mitigate risks, prevent accidents, and ensure compliance with safety regulations, ultimately fostering a culture of safety and wellbeing on construction sites.

The integration of AI for predictive maintenance and asset management represents a change in thinking in the construction industry, offering unparalleled insights, efficiencies, and safety enhancements. As construction companies continue to embrace AI technology, they are poised to unlock new levels of operational excellence, cost savings, and project success in an increasingly competitive and complex industry landscape. The transformative impact of AI on maintenance and asset management practices underscores its potential to revolutionize construction operations and drive sustainable growth in the digital era.

Smart Construction Site Monitoring and Progress Tracking

In the rapidly evolving landscape of the construction industry, the convergence of innovative technologies has revolutionized the way project monitoring and progress tracking are conducted. The seamless integration of artificial intelligence (AI), Internet of Things (IoT), drones, and advanced data analytics has ushered in a new era of intelligence-driven construction site management.

At the heart of smart construction site monitoring lies the sophisticated network of IoT devices that are strategically deployed throughout the construction site. These sensors, capable of capturing real-time data on a multitude of environmental and operational parameters, provide project managers with a comprehensive understanding of the site dynamics. From monitoring temperature and humidity levels to tracking noise emissions and detecting safety hazards, IoT devices serve as the eyes and ears of the construction project, facilitating proactive decision-making and risk mitigation.

Drones have emerged as indispensable assets in the arsenal of tools for construction site monitoring. Equipped with high-resolution cameras, thermal imaging capabilities, and LiDAR sensors, drones offer unparalleled visibility from aerial vantage points. By conducting regular flyovers and capturing detailed images and videos of the site, drones enable project managers to assess progress, identify bottlenecks, and spot deviations from the original plan with remarkable precision. The ability to rapidly survey large areas and generate 3D models of the site enhances communication, collaboration, and problem-solving among project stakeholders.

The fusion of building information modeling (BIM) with AI-powered algorithms represents a transformative leap in progress tracking and analysis. By overlaying BIM models with real-time data streams from IoT sensors and drone imagery, project managers can create a dynamic digital twin of the construction site. This virtual representation allows for instant comparisons between the planned schedule and the actual progress on the ground, facilitating predictive analytics, resource optimization, and early identification of potential delays or schedule deviations. The constructive interaction of BIM and

AI enables project managers to make data-driven decisions, optimize workflows, and drive continuous improvement throughout the construction lifecycle.

In conclusion, the adoption of smart construction site monitoring and progress tracking technologies is reshaping the industry by enhancing efficiency, transparency, and accountability. By harnessing the power of AI, IoT, drones, and data analytics, construction companies can unlock new levels of operational excellence, deliver projects on time and within budget, and pave the way for a more sustainable and resilient built environment.

Future Trends and Innovations in AI for Construction Industry

As the construction industry continues to evolve, artificial intelligence (AI) technologies are playing an increasingly vital role in shaping its future landscape. These innovations are driving advancements in efficiency, safety, and project management, revolutionizing the way construction projects are executed.

One of the prominent trends in the industry is the use of drones equipped with AI capabilities. These aerial vehicles are transforming traditional construction practices by providing real-time data for surveying, mapping, and monitoring construction sites. With advanced imaging and data processing capabilities, drones can capture detailed information, detect issues early on, and enhance decision-making throughout the project lifecycle.

Moreover, the integration of virtual and augmented reality technologies is revolutionizing project visualization and design processes. By creating immersive, 3D models of construction projects, stakeholders can collaborate more effectively, identify potential design flaws, and optimize layouts for improved efficiency. AI algorithms analyze data from these technologies to generate insights that drive informed decision-making and streamline project workflows.

Machine learning algorithms are increasingly used for predictive modeling and risk assessment in construction projects. By leveraging historical project data and real-time information, AI systems can forecast potential risks, delays, and resource requirements. This predictive capability enables project managers to proactively address issues, optimize schedules, and minimize disruptions, ultimately enhancing project success rates and profitability.

The proliferation of Internet of Things (IoT) devices in construction is also reshaping project monitoring and management practices. These connected devices collect data on equipment performance, material inventory, and environmental conditions, enabling real-time tracking and analysis. AI algorithms process this data to identify patterns, trends, and anomalies, empowering stakeholders to make data-driven decisions that enhance operational efficiency and sustainability.

Furthermore, the development of autonomous construction vehicles and robots is transforming traditional construction practices. These machines can perform repetitive or hazardous tasks with precision and speed, enhancing productivity and worker safety. By leveraging AI-powered automation, construction companies can optimize workflows, reduce labor costs, and deliver projects more efficiently.

In conclusion, the future of AI in the construction industry is characterized by a convergence of innovative technologies that promise to optimize project delivery, enhance safety standards, and drive sustainable practices. By embracing these advancements, construction companies can position themselves at the forefront of industry transformation, unlocking new opportunities for growth, efficiency, and competitiveness.

Beyond the technology itself, the implementation of AI in construction also brings about significant shifts in workforce dynamics. As AI systems take on more tasks traditionally done by human workers, there is a growing need for upskilling and reskilling to ensure that employees can effectively collaborate with these technologies. This transformation in the workforce requires initiative-taking measures from companies to invest in training programs and create a culture of continuous learning to adapt to the evolving industry landscape.

Moreover, the ethical implications of AI in construction cannot be overlooked. Issues surrounding data privacy, algorithm bias, and job displacement require careful consideration and proactive measures to safeguard the interests of all stakeholders involved. Regulatory frameworks and industry standards must be established to ensure the responsible deployment and use of AI technologies in construction projects, fostering trust and transparency within the industry.

The potential of AI in construction goes beyond improving operational efficiency; it also holds promise for driving sustainability initiatives within the built environment. By harnessing AI to optimize resource utilization, reduce waste, and enhance energy efficiency, construction companies can contribute to creating more environmentally friendly and resilient structures. This shift towards sustainable practices aligns with global efforts to combat climate change and build a more sustainable future for the built environment.

In essence, the integration of AI technologies in construction represents a transformative journey towards a more efficient, collaborative, and sustainable industry. By harnessing the power of AI alongside human expertise, construction companies can unlock new possibilities for innovation, growth, and positive impact on the built environment and society.